Fernanda Brum Pires
Marcelo Barcellos da Rosa

Constituents of some medicinal species from the Brazilian flora

Fernanda Brum Pires
Marcelo Barcellos da Rosa

Constituents of some medicinal species from the Brazilian flora

An approach to phenolic compounds

ScienciaScripts

Imprint
Any brand names and product names mentioned in this book are subject to trademark, brand or patent protection and are trademarks or registered trademarks of their respective holders. The use of brand names, product names, common names, trade names, product descriptions etc. even without a particular marking in this work is in no way to be construed to mean that such names may be regarded as unrestricted in respect of trademark and brand protection legislation and could thus be used by anyone.

Cover image: www.ingimage.com

This book is a translation from the original published under ISBN 978-3-330-76207-7.

Publisher:
Sciencia Scripts
is a trademark of
Dodo Books Indian Ocean Ltd. and OmniScriptum S.R.L publishing group

120 High Road, East Finchley, London, N2 9ED, United Kingdom
Str. Armeneasca 28/1, office 1, Chisinau MD-2012, Republic of Moldova, Europe
Managing Directors: Ieva Konstantinova, Victoria Ursu
info@omniscriptum.com

Printed at: see last page
ISBN: 978-620-8-38291-9

AUTHOR DETAILS

Carolina Bolssoni Dolwitsch

Postgraduate student (PhD) at the Postgraduate Programme in Pharmaceutical Sciences at the Federal University of Santa Maria (UFSM). Degree in Industrial Chemistry from UFSM. Undergraduate student in Pharmacy at UFSM.

Fernanda Brum Pires

Postgraduate student (PhD) at the Postgraduate Programme in Pharmaceutical Sciences at the Federal University of Santa Maria (UFSM). Specialist in Pharmacology and Drug Interactions UNINTER (2014). Degree in Pharmacy with generalist training from the Franciscan University Centre - UNIFRA.

Marcelo Barcellos da Rosa (ORGANISER)

Professor at the Federal University of Santa Maria (UFSM). Post-doctorate at the Bayreuth Centre of Ecology and Environmental Research, Bayreuth - Germany and at the National Institute for Space Research (INPE) - Brazil. PhD in Natural Sciences from the Gottfried Wilhelm Leibniz Universitat Hannover - Germany. Master's Degree in Chemistry from UFSM and Bachelor's Degree in Industrial Chemistry and Chemistry from UFSM.

CONTENTS

CHAPTER 1

THE SEARCH FOR NATURAL ACTIVE INGREDIENTS

Fernanda Brum Pires

Carolina Bolssoni Dolwitsch

Marcelo Barcellos da Rosa

Since ancient times, drugs from nature have been used as a therapeutic resource to treat diseases (PETROVSKA, 2012; SCHMITZ et al., 2005). The use of medicinal plants is an ancient practice (VEIGA; PINTO; MACIEL, 2005), which represents part of the history of our ancestors on earth (REZENDE; COCCO, 2002). As we know, in the early days of mankind there was no information about the reasons for illnesses, nor was it even known how plants could cure ailments, so this activity took place instinctively and based on experience (PETROVSKA, 2012). The use of phytotherapy originated from popular knowledge and for a long time remained the only alternative for health problems (ALVIM et al., 2006).

Over the years, the use of plants has gradually abandoned the empirical side as the efficacy of certain species has been scientifically proven (PETROVSKA, 2012). In recent decades, it has been consolidated as a complement to modern medicine (ALVIM et al., 2006).

Brazil is a country with great potential for research involving plant species, as it has the greatest biodiversity on the planet (PORT'S et al., 2013). Among the various existing biomes, the Amazon Rainforest stands out the most, since it alone comprises a quarter of the world's animal and plant species (BIESKI et al., 2015).

However, the majority of Brazilian plant species have not yet been studied, representing a potential economy that needs to be exploited (BOLZANI et al., 2012). In this sense, studies involving plant species are needed to identify their chemical constituents in order to justify their pharmacological activities.

Plant metabolism is known to occur through the synthesis of primary and secondary metabolites. Secondary metabolism plays an important role in the process of plant adaptation to the environment and also represents a source of bioactive products (BOURGAUD et al., 2001).

The medicinal properties of plants are attributed to the presence of secondary metabolites in their constitution (PAREKH; KARATHIA; CHANDA, 2006), including phenolic compounds, a widely distributed group in the plant kingdom responsible for various therapeutic properties (CARVALHO; GOSMANN; SCHENKEL, 2004, p. 519, 528; PIETTA, 2000).

Numerous species of Brazilian flora are of medicinal interest. These include Connarus perrottetti var. angustifolius, Cecropia palmata, Cecropia obtusa and Mansoa alliacea, which are native to the Amazon and are widely used by the local population for various health problems.

Considering the characteristics of the Amazon region, the climate is quite peculiar since it comprises a "dry period" from July to October and a "rainy period" from December to May (ANANIAS et al., 2010). Species can respond differently, since the content of secondary metabolites is largely influenced by environmental factors (GOBBO-NETO; LOPES, 2007).

In recent years, there has been a growing interest in polyphenols in vegetables. Both for research and human consumption, due to their nutritional potential and added therapeutic value (AJILA et al., 2011).

In view of the scant information in the literature about these species, and also considering that no studies have been found dedicated to evaluating their chemical constitution, especially with regard to evaluation during different sampling periods, research such as this is relevant. This sparked an interest in obtaining consistent scientific data on the species in the light of the theme, both in terms of phenolic constitution and behaviour over different collection periods, while also seeking to establish a relationship between the bioactive compounds identified and the activities described by folk medicine.

Research into natural products combined with analytical methods involving high-performance liquid chromatography and mass spectrometry make it possible to identify, quantify and structurally confirm the compounds of interest (BOLZANI et al., 2012). Consequently, they help in the investigation of the constituents responsible for medicinal actions.

The plants that are the subject of this study are being researched in conjunction with other research groups, which allows for an interdisciplinary study. The species were chosen for their broad pharmacological action as well as their nutritional appeal, both of which are mentioned in popular medicine.

This work aims to study the chemical constitution of the species Connarus perrottetti var. angustifolius, Cecropia palmata, Cecropia obtusa and Mansoa alliacea, considering the chromatographic profile in extracts of different polarities obtained in three years of collection.

CHAPTER 2

OVERVIEW OF MEDICINAL PLANTS

Fernanda Brum Pires

Carolina Bolssoni Dolwitsch

Marcelo Barcellos da Rosa

2.1 PLANT METABOLISM

Metabolism is the set of transformations of organic molecules, catalysed by enzymes, which occur in living cells, providing energy and guaranteeing the continuity of the species. Plant metabolism comprises metabolites, known as primary and secondary (PEREIRA; CARDOSO, 2012). Primary or essential metabolism is responsible for synthesising cellulose, lignins, proteins, lipids, sugars and other substances related to vital functions (PEREIRA; CARDOSO, 2012), such as cell division and growth, respiration, storage and reproduction (BOURGAUD et al., 2001).

The secondary or non-essential metabolism is produced from the primary metabolism, by amino acids or acetyl coenzyme A (MATSUMURA et al., 2013, p 4). This is directly involved in the mechanisms that allow plants to adapt to the environment (PEREIRA; CARDOSO, 2012), as it provides protection against pathogens such as insects, fungi, viruses and bacteria and ultraviolet radiation (SANTOS, 2004, p 404; TIVERON et al., 2012). They are also related to the fact that they confer various biological activities and are commercially important due to their pharmacological interest (BOURGAUD et al., 2001; ZUANAZZI; MONTANHA, 2004, p 601) and nutritional value (HUBER; AMAYA, 2008; PEREIRA; CARDOSO, 2012).

Secondary metabolites are classified based on their biosynthetic origin, resulting in various classes of bioactive compounds (SANTOS, 2004, p 411). According to

Pereira and Cardoso (2012), their origin can be summarised from the metabolism of glucose, through two main intermediates: shikimic acid and acetate.

Shikimic acid is the precursor of constituents such as hydrolysable tannins, coumarins, some alkaloids and phenylpropanoids, while acetate derivatives give rise to terpenoids, steroids, condensed tannins and some alkaloids, among others, as shown in figure 1:

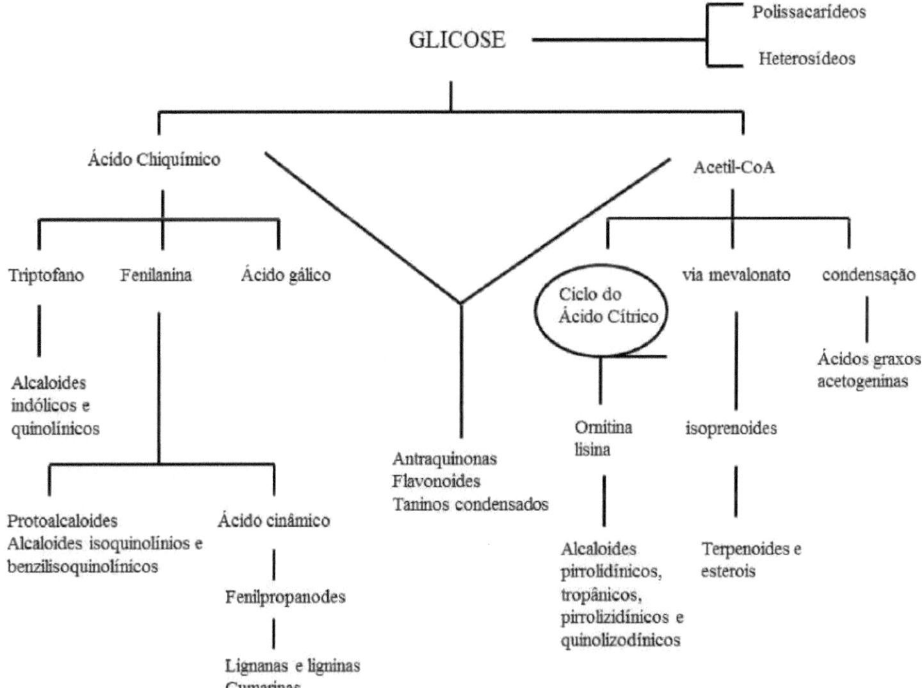

Figure 1 - Biosynthesis of secondary metabolites:

Source: (SANTOS, 2004, p 411).

2.2 PHENOLIC COMPOUNDS

The term "phenolic" or "polyphenolic" can be defined as a substance that has one or more aromatic nuclei containing hydroxylated substituents and/or their functional derivatives (esters, methyl ethers, glycosides and others). However, its biosynthetic origin must also be taken into account for a coherent definition (ZUANAZZI; MONTANHA, 2004, p 577).

Phenolic compounds are widely distributed in the plant kingdom (CARVALHO; GOSMANN; SCHENKEL, 2004, p. 519) and can range from simple phenolics to highly polymerised structures (NACZK; SHAHIDI, 2004). They are divided into at least 10 classes according to their basic chemical structure (BRAVO, 1998), and Table 1 shows the compounds of interest in this work.

Table 1 - Phenolic compounds studied:

CLASSES/ COMPOUNDS PHENOLICS	**STRUCTURAL FORMULA**
Hydroxybenzoic acid Gallic acid	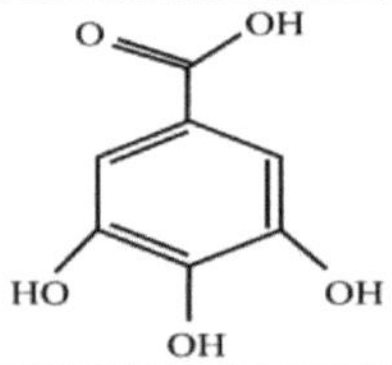
Hydroxycinnamic acid Caffeic acid: R=OH Ferulic acid: R=OCH3	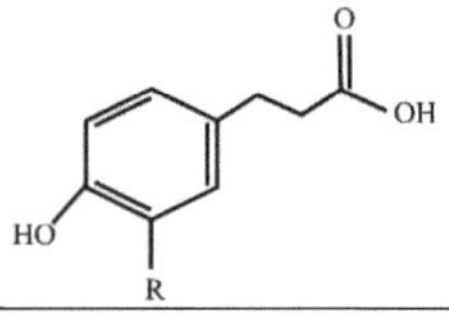
Stilbene Resveratrol	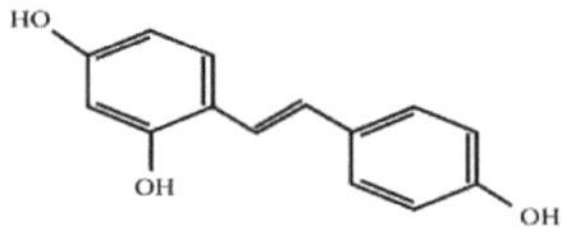
Flavanol Catechin	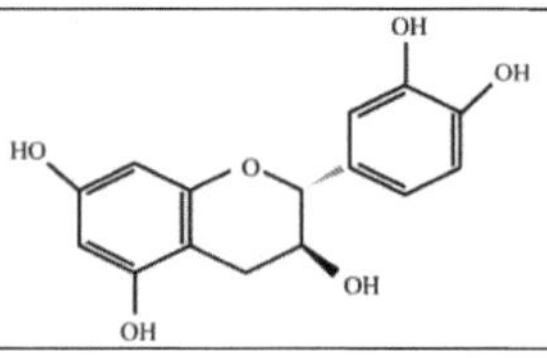
Flavone Chrysin: R1 = R2 = OH Flavone: R1 = R2 = H	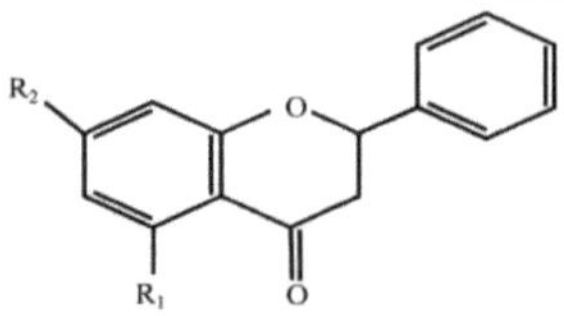

Flavonol - glycosylated
Quercitrin: R=raminosis
Rutin: R=rutinose

Phenolic acids come from the shikimic acid pathway and can originate from benzoic acid (gallic acid), which has a basic skeleton (C6-C1), while others originate from hydroxycinnamic acid (caffeic and ferulic acid), with a common skeleton (C6-C3). The combination of shikimic acid units with acetate units, with a basic structure (C6-C2-C6), gives rise to stilbenes and flavonoids (C6-C3-C6), the latter of which can be subdivided into various classes depending on the substituents on the nuclei (BRAVO, 1998; SANTOS, 2004, p 411-412, 422).

2.3 MEDICINAL PLANTS

Even with the great advances in modern medicine, phytotherapy has a very significant place. According to the World Health Organisation, around 80% of the population in developing countries use plants or preparations thereof as a basic therapeutic resource (BRASIL, 2006). It is estimated that around 30 per cent of the drugs available on the world market come from natural sources (IPEA, 2011).

In this way, the data referenced demonstrates the importance of research aimed at identifying chemical constituents, in this case phenolic compounds, since they contribute scientific information about the use of the species in traditional medicine and because of the numerous therapeutic properties they can offer.

Table 1 shows the botanical characteristics, some phytocompounds already identified and the recognised use in folk medicine of the genera of interest in the study.

Table 1 - General overview of the *Connarus, Cecropia and Mansoa* genera.

GENUS - SPECIES	FEATURES	PHYTOCOMPOUNDS	USE POPULAR
Connarus perrottetti var. Angustifolius	The genus Connarus belongs to the Connaraceae family, has 16 genera and around 300 species, distributed in tropical regions. Popularly known as Pará barbatimão (PARACAMPO, 2011).	Tannins, catechins, flavonoids, cardiotonic glycosides, coumarinase saponins (OLIVERIA 2013).	Treatment of genitourinary infections, uterine haemorrhage, vaginal discharge, headaches, gastric diseases and coughs (COELHO-FERREIRA, 2009).
Cecropia obtusa *Cecropia palm*	The genus Cecropia belongs to the Moraceae/Cecropiacea family and has around 75 species distributed in Brazil. The trees have straight, hollow trunks that reach 4 to 30 metres and large leaves. They are popularly known as embaúbas (ARAGÃO et al., 2013; BERINGHS et	Flavonoids, catechins, triterpenes, tannins, steroids and saponins (AREND, 2010; ROCHA et al., 2007; SANTOS et al., 2011).	Tr eatment of respiratory diseases (cough , asthma, expectorant) hypertension, anti-inflammatory, antimicrobial and

	al., 2015; CAICEDO, 2005).		diuretic (AREND, 2010; FREITAS; FERNANDES, 2006; SANTOS etal., 2011).
Mansoa alliacea . i vtv k	The genus Mansoa belongs to the Bignoniaceae family and includes 11 species distributed mainly in dry and humid forests. Popularly known as garlic liana (RIBEIRO 2008; ZOGHBI et al., 2009).	Alkaloids, tannins, flavonoids, quinone glycosides, steroids, coumarins and derivativ es (OLIVEIRA 2013; ZOGHBI et al., 2009).	Tr eatment of colds, pneumonia and malaria, as an insecticide and antirheumatic (PATEL et al., 2013; ZOGHBI et al., 2009).

2.4 EXTRACTION METHODS

Extraction is the main procedure for recovering and isolating bioactive compounds from plant materials prior to analysis (STALIKAS, 2007). Among the various extraction methods, there are the so-called conventional ones, such as soxhlet, hydrodistillation and maceration, and those considered alternative to these, such as ultrasound, supercritical fluid, among others (VINATORU, 2001; WU et al., 2015).

In recent years, in order to overcome the disadvantages of conventional extraction methods, there has been a growing interest in alternative methods, such as ultrasound-assisted extraction (UAE). The extraction process has advantages such as simplicity of operation, reduced process time, less solvent used and a higher final yield (PORTO; PORRETTO; DECORTI, 2013).

(EAU) occurs based on the phenomenon of acoustic cavitation, which refers to the formation, growth and implosion of gas bubbles. The collapse of cavitation bubbles in the liquid phase near the plant cell wall promotes cell rupture, facilitating solvent penetration and intensifying mass transfer (AJILA et al., 2011; BENELLI, 2010; MA et al., 2008).

Studies by Hojnik et al. (2007) evaluated different techniques for extracting compounds from Urtiga dioica, which showed that ultrasound is a promising alternative method since it allowed for greater extraction efficiency. This method has increased the effectiveness of extractions involving polyphenols in plant matrices, as it reduces extraction time and does not cause degradation of thermolabile compounds (AJILA et al., 2011). It is recognised for its potential application in the phytopharmaceutical industry (VILKHU et al., 2008).

It is known that there is no satisfactory extraction method for isolating phenolic compounds or a specific class, since extraction is influenced by various factors such as: chemical nature, particle size, time and storage conditions. In addition, their solubility is influenced by the

polarity of the solvent, making it necessary to use different solvents and combinations with water (NACZK; SHAHIDI, 2004, 2006; STALIKAS, 2007), and may also be complexed with other plant components and other non-phenolic substances, affecting extraction procedures (AJILA et al., 2011).

Like other methods, this one also has its limitations, since ultrasound bath-assisted extraction loses out in terms of efficiency compared to the ultrasound probe (LUQUE-GARCIA; LUQUE DE CASTRO, 2003). However, the extraction methodology used reflects what you want to obtain, as well as the instrumentation available.

2.5 CHARACTERISATION METHODS

There are various methods reported in the literature for obtaining, detecting and quantifying phenolic compounds in plant extracts, ranging from the most preliminary such as phytochemical analysis to more sophisticated methodologies such as liquid chromatography coupled with mass detection (LC-MS), gas chromatography coupled with mass detection (GC-MS) and nuclear magnetic resonance (NMR).

Phytochemical analysis tests are classic tools, based on carrying out chemical reactions that result in the development of colouring and/or precipitates characteristic of groups of substances (FALKENBERG; SANTOS; SIMÕES, 2004, p 234-235).

According to Ignat and colleagues (2013), high-performance liquid chromatography (HPLC) is the most widely used technique for separating and characterising polyphenols. It has versatile and adaptable instrumentation with advantages such as high selectivity, sensitivity, resolution, precision and sample preservation.

However, ultraviolet/visible (UV/VIS) detection based solely on comparing the retention times of compounds with analytical standards can lead to identification errors when dealing with complex matrices such as plant extracts. In this context, LC-MS hyphenation techniques overcome the aforementioned problem, as they utilise the separation potential of HPLC and mass structural elucidation (IGNAT; VOLF; POPA, 2013, p 2077).

Gas chromatography is one of the main techniques available for analysing phenolic acids in plants. However, due to the hydroxyl group in their structure, they have low volatility, requiring chemical derivatisation, a process that converts hydroxyl groups into ethers or esters, making these compounds more volatile and able to be identified using this methodology. Its association with mass spectrometry guarantees sensitivity and specificity in analyses (STALIKAS, 2007).

Nuclear magnetic resonance is a complementary technique used in the structural elucidation of phenolic constituents isolated from different medicinal plants. It requires a large amount of sample to analyse, but when it is two-dimensional it does not need reference standards and has several advantages over other instruments, being more precise and accurate than HPLC. It also provides information on impurities and structures of isomeric compounds that are impossible to differentiate by LC-MS (AJILA et al., 2011).

Table 2 contains information found in the literature on the extraction, characterisation and phytocompounds identified in relation to the genera under study.

Table 2 - Literature on obtaining and characterising phenolic constituents for species of the Connarus, Cecropia and Mansoa genera.

PLANT	EXTRACTION METHOD	CHARACTERISATION METHOD	COMPOUNDS IDENTIFIED
Connarus perrottetti var. angustifolius	Under reflux	Phytochemical analysis	Saponins, coumarins and flavonoids (OLIVEIRA, 2013).
	Maceration	HPLC-PAD	Gallic acid, caffeic acid and catechin (DA SILVEIRA et al., 2015).
Cecropia glaziovii	Maceration and percolation	HPLC-UV	Rutin, quercetin, epicatechin and catechin (CAICEDO, 2005).
C. glaziovii	Maceration	HPLC-UV	Caffeic and chlorogenic acid (BERINGHS et al., 2015).
C. glaziovii and pachystachya	Infusion	HPLC-DAD	Chlorogenic acid, isoorientin, orientin and isovitexin (COSTA et al., 2011).
Cecropia obtusa and palmata	Maceration	HPLC-PAD	P-coumaric acid, ferulic acid, caffeic acid, resveratrol and catechin (DA SILVEIRA et al., 2015).

Mansoa alliacea	Maceration	Phytochemical analysis	Tannins, flavonoids and catechin (PATEL et al., 2013).Triterpenes and alkaloids (RIBEIRO, 2008).
	Maceration	HPLC-PAD	P-coumaric acid, ferulic acid and resveratrol (DA SILVEIRA et al., 2015).
	Under reflux	Phytochemical analysis	Alkaloids and tannins (OLIVEIRA, 2013).

The techniques used in this work to characterise, quantify and confirm the structure of the species are listed below:

High performance liquid chromatography with diode array detector (HPLC-DAD) provides the UV spectral data of the separated compounds; thus, when a peak matches the retention time of one of the analytical standards, it can be identified by spectral comparison of both (IGNAT; VOLF; POPA, 2013 p 2076).

Liquid chromatography coupled with sequential mass spectrometry (LC-MS/MS) is widely used to avoid the ambiguities associated with other techniques. In this sense, the coupling of two or more mass analysers to the system further increases sensitivity and selectivity in detection, as they generate additional structural data relevant to identification. The ionisation sources generally used to characterise phenolic compounds are APCI and ESI "atmospheric pressure chemical ionisation" and "electrospray ionisation" respectively (IGNAT; VOLF; POPA, 2013 p 2080, 2082).

In the literature, several studies have demonstrated the importance and efficiency of HPLC-DAD and LC-MS/MS methods in the separation, identification and quantification of polyphenols in plant species, which is why they were used to achieve the objectives of this work.

2.6 MEDICINAL PROPERTIES OF POLYPHENOLS

Phenolic constituents are recognised for their various medicinal properties, such as antioxidant, antibacterial, antiviral (ZUANAZZI; MONTANHA, 2004, p 601-606), anti-inflammatory and anti-allergic (PIETTA, 2000), among others. Below is a brief mention of the main pharmacological activities described in relation to phenolic phytocompounds.

2.6.1 Antioxidant activity

The best-known benefit of polyphenols for human health is attributed to their antioxidant properties (BANERJEE; RAJAMANI, 2013). According to Yao et al. (2004), these compounds have an ideal chemical structure for free radical scavenging activity, since they not only have phenolic hydroxyl groups that are prone to donating a hydrogen atom or an electron to a free radical, but also an extended conjugated aromatic system to displace an unpaired electron.

The antioxidant potential of phenolic compounds is related to the number and arrangement of hydroxyl groups in the molecule of interest (SHAHIDI; AMBIGAIPALAN, 2015). The antioxidant activity of flavonoids depends on their chemical structure; in general, the greater the number of hydroxyls, the greater the activity as a hydrogen and electron donor agent (ALVES et al., 2007).

Derivatives of phenolic acids such as caffeic and ferulic acid, quercetin, kaempferol, myricetin, among others, are reported in the literature to have antioxidant activity (CARVALHO; GOSMANN; SCHENKEL, 2004, p. 528; ZUANAZZI; MONTANHA, 2004, p. 602). In view of this, these compounds can act by delaying diseases related to oxidative stress (CARVALHO; GOSMANN; SCHENKEL, 2004, p. 528), such as diabetes, cancer (SILVA; JASIULIONIS, 2014) and rheumatism (FERREIRA; MATSUBARA, 1997), so the presence of these constituents in plant species justifies their therapeutic use for such purposes.

2.6.2 Anti-inflammatory activity

In terms of anti-inflammatory activity, flavonoids act by modulating cells involved in inflammation (inhibiting the proliferation of T lymphocytes), inhibiting the production of pro-inflammatory cytokines (TNF-α and IL-1), modulating the activity of arachidonic acid pathway enzymes such as phospholipase A2, cyclooxygenase and lipooxygenase, as well as modulating the nitric oxide-forming enzyme. Various compounds such as quercetin, kaempferol, rutin, apigenin, luteolin, chrysin, vitexin (COUTINHO et al., 2009) and resveratrol (ANTUS et al., 2015; LÓPEZ- POSADAS et al., 2008) have been mentioned for their anti-inflammatory properties, thus supporting the medicinal use of species containing them in search of therapeutic properties.

2.6.3 Antimicrobial activity

Compounds of natural origin have increasingly aroused interest in research, especially in terms of exploring antimicrobial activity, due to the increase in cases of bacterial resistance as a result of the misuse of antibiotics. In this sense, studies involving phenolic compounds have been highlighted (BORGES et al., 2013).

For polyphenols, just as the site and number of hydroxyl groups influence antioxidant capacity, they are also related to toxicity to microorganisms, with an increase in hydroxyls resulting in increased toxicity. The action of compounds such as gallic and ferulic acid involves mechanisms such as the rupture and formation of pores in bacterial cell membranes (BORGES et al., 2013) which have been demonstrated for both positive and negative bacteria.

In addition, gallic acid, catechin (DÍAZ-GÓMEZ et al., 2013, 2014; SARJIT et al., 2015) and resveratrol (BRONW et al., 2009) are reported in the literature as antibacterial agents.

Studies involving antiviral activity mainly include flavonoids such as quercetin and its derivatives for inhibiting viral replication. Quercitrin, kaempferol, apigenin, chrysin, luteolin and galangin have also been described as antiviral agents (ZUANAZZI; MONTANHA, 2004, p 601-602, 607).

According to Zhang and colleagues (2013), rutin, apigenin and quercetin have antifungal activity. This justifies the therapeutic use of species with these constituents in order to treat viral, fungal and bacterial infections.

REFERENCES

AJILA, C. M. et al. Extraction and Analysis of Polyphenols: Recent Trends. **Critical Reviews in Biotechnology**, v. 31, n. 3, p. 227-249, 2011.

ALVES, C, Q. et al. Evaluation of the antioxidant activity of flavonoids. **Dialogues and Science**, n. 12, p. 1-8, 2007.

ALVIM, N. A. T. et al. The use of medicinal plants as a therapeutic resource: from the influences of professional training to the ethical and legal implications of its applicability as an extension of the nurse's caring practice. **Revista Latino-Americana de Enfermagem**, v.14, n. 3, p. 316-323, 2006.

ANANIAS, D.'dos S. et al. Climatology of the vertical structure of the atmosphere in November for Belém-PA. **Revista Brasileira de Meteorologia**, v. 25, n. 2, 218226, 2010.

ANTUS, C. et al. Anti-inflammatory effects of a triple-bond resveratrol analog: Structure and function relationship. **European Journal of Pharmacology**, v. 748, n. 4, p. 61-67, 2015.

ARAGÃO, D. M. de O. et al. Anti-Inflammatory, antinociceptive and cytotoxic effects of the methanol extract of Cecropia pachystachya Trécul. **Phytotherapy Research**, v. 27, n. 6, p. 926-930, 2013.

AREND, D. P. **Development of a microstructured system containing a standardised extract of *Cecropia glaziovii* Sneth (Embaúba)**. 2010. 198 p. Dissertation (Master's in Pharmacy) - Federal University of Santa Catarina, Florianópolis, SC, 2010.

BANERJEE, S.; RAJAMANI, P. **Cellular, Molecular, and Biological Perspective of Polyphenols in Chemoprevention and Therapeutic Adjunct in Cancer**. In: RAMAWAT, K. G.; MÉRILLON, J-M. Natural Products Phytochemistry, Botany and Metabolism of Alkaloids, Phenolics and Terpenes; Berlin: Springer, 2013. chap.

71, p. 2184.

BENELLI, P. **Adding value to orange pomace (*Citrus sinensis* L. Osbeck) by obtaining bioactive extracts using different extraction techniques**. 2010. 233 p. Dissertation (Master's in Food Engineering) - Federal University of Santa Catarina, Florianópolis, SC, 2010.

BERINGHS, A. O. et al. Response Surface Methodology IV-Optimal design applied to the performance improvement of an RP-HPLC-UV method for the quantification of phenolic acids in Cecropia glaziovii products. **Revista Brasileira de Farmacognosia**, v. 25, n. 5, p. 513-521, 2015.

BIESKI, I. G. C. et al. Ethnobotanical study of medicinal plants by population of Valley of Juruena Region, Legal Amazon, Mato Grosso, Brazil. **Journal of Ethnopharmacology**, v. 15, n. 173, p. 383-423, 2015.

BOLZANI, V. da S. et al. Natural products from Brazilian biodiversity as a source of new models for medicinal chemistry. **International Union of Pure and Applied Chemistry**, v. 84, n. 9, p. 1837-1846, 2012.

BORGES, A. et al. Antibacterial Activity and Mode of Action of Ferulic and Gallic Acids Against Pathogenic Bacteria. **Microbial drug resistance**, v. 19, n. 4, p. 256265, 2013.

BOURGAUD, F. et al. Production of plant secondary metabolites: a historical perspective. **Plant Science**, v. 161, n. 5, p. 839-851, 2001.

BRAZIL. Ministry of Health. **National Policy on Medicinal Plants and Herbal Medicines**. Brasília, 2006. Available at: <http://bvsms.saude.gov.br/bvs/publicacoes/politica_nacional_fitoterapicos.pdf> Accessed on: 22 Dec. 2015.

BRAVO, L. Polyphenols: Chemistry, Dietary Sources, Metabolism, and Nutritional Significance. **Nutrition Reviews**, v. 56, n. 11, p. 317-333, 1998.

BROWN, J. et al. Antibacterial effects of grape extracts on Helicobacter pylori. **Applied and Environmental Microbiology**, v. 75, n. 3, p. 848-852, 2009.

CAICEDO, P. E. L. **Contribution to the standardisation of** Cecropia glaziovii **Snethl. leaf extracts: Studies of seasonal and intraspecific variation in flavonoids and proanthocyanidins, extraction methodologies and vasorelaxant activity**. 2005. 292 f. Thesis (Doctorate in Pharmaceutical Sciences) - Universidade Federal de Minas Gerais, Belo Horizonte, MG, 2005.

CARVALHO, J. C.T.; GOSMANN. G.; SCHENKEL, E. P. **Simple phenolic compounds and heterosides**. In: SIMÕES, C. M. O. et al. Farmacognosia - da Planta ao Medicamento. 5. ed. Porto Alegre: Editora da UFRG/Editora da UFSC, 2004. chap. 20, p. 519, 528.

COELHO-FERREIRA, M. Medicinal knowledge and plant utilisation in an Amazonian coastal community of Maruda, Pará State (Brazil). **Journal of Ethopharmacology**, v. 126, n. 1, p. 159-175, 2009.

COSTA, G. M. et al. An HPLC-DAD Method to Quantification of Main Phenolic Compounds from leaves of Cecropia species. **Journal of the Brazilian Chemical Society**, v. 22, n. 6, p. 1096-1102, 2011.

COUTINHO, M. A. S.; MUZITANO, M. F.; COSTA, S. S. Flavonoids: Potential Therapeutic Agents for the Inflammatory Process. **Revista Virtual de Química**, v.1, n. 3, p. 241-256, 2009.

DÍAZ-GÓMEZ, R. et al. Combined effect of gallic acid and catechin against Escherichia coli. **Food Science and Technology**, v. 59, p. 896-900, 2014.

DÍAZ-GÓMEZ, R. et al. Comparative antibacterial effect of gallic acid and catechin against Helicobacter pylori. **Food Science and Technology**, v. 54, p. 331-335, 2013.

FALKENBERG, M. B.; SANTOS, R. I. **Introduction to phytochemical analysis**. In: SIMÕES, C. M. O. et al. Farmacognosia - da Planta ao Medicamento. 5. ed. Porto Alegre: Editora da UFRG/Editora da UFSC, 2004. chap. 10, p. 234-235.

FERREIRA, A. L. A.; MATSUBARA, L. S. Free radicals: Concepts, related diseases, defence system and oxidative stress. **Revista da Associação Médica Brasileira**, v. 43, n. 1, p. 61-68, 1997.

FREITAS, J. C.; FERNANDES, M. E. B. Use of medicinal plants by the community of Enfarrusca, Bragança, Pará. **Boletim do Museu Paraense Emílio**

Goeldi, Ciências Naturais, Belém, v. 1, n. 3, p. 11-26, 2006.

GOBBO-NETO, G. L.; LOPES, N. P. Medicinal plants: factors influencing the content of secondary metabolites. **Química Nova**, v. 30, n. 2, p. 374-381, 2007.

HOJNIK, M.; SKERGET, M.; KNEZ, Z. Isolation of chlorophylls from stinging nettle (Urtica dioica L.). **Separation and Purification Technology**, n. 57, p. 37-46, 2007.

HUBER, L. S.; AMAYA, D. B. R. Flavonols and flavones: Brazilian sources and factors influencing food composition, **Alimentos e Nutrição**, v. 19, n. 1, p. 97108, 2008.

IGNAT, I.; VOLF, I.; POPA, V. I. **Analytical Methods of Phenolic Compounds**. In: RAMAWAT, K. G.; MÉRILLON, J-M. Natural Products Phytochemistry, Botany and Metabolism of Alkaloids, Phenolics and Terpenes; Berlin: Springer, 2013. chap. 67, p. 2076-2077, 2080, 2082.

INSTITUTO DE PESQUISA ECONÔMICA APLICADA, **Environmental sustainability in Brazil: biodiversity, economy and human well-being.** Brasilia, 2010. Available at: <http://www.ipea.gov.br/portal/images/stories/PDFs/livros/livros/livro07_sustentabilidadeambienta.pdf > Accessed on: 10 Dec. 2015.

LUQUE-GARCIA, J. L.; LUQUE DE CASTRO, M. D.; Ultrasound: a powerful tool for leaching. **Trends in Analytical Chemistry**, v. 22, n. 1, p. 41-47, 2003.

LUQUE DE CASTRO, M. D.; PRIEGO-CAPOTE, F. Ultrasound-assisted preparation of liquid samples. **Talanta**, v. 72, n. 2, p. 321-334, 2007.

MA, Y. et al. Phenolic compounds and antioxidant activity of extracts from ultrasonic treatment of Sastsuma Mandarin (Citrus unshiu Marc.) Peels. **Journal Agricultural Food Chemistry**, v. 56, n. 14, p. 5682-90, 2008.

MATSUMURA, E. et al. **Microbial Production of Plant Benzylisoquinoline Alkaloids**. In: RAMAWAT, K. G.; MÉRILLON, J-M. Natural Products Phytochemistry, Botany and Metabolism of Alkaloids, Phenolics and Terpenes; Berlin, Heidelberg: Springer, 2013. ch. 1, p. 4.

NACZK, M.; SHAHIDI, F. Extraction and analysis of phenolics in food. **Journal of Chromatography A**, v. 1054, n. 1, p. 95-111, 2004.

LÓPEZ-POSADAS, R. et al. Effect of flavonoids on rat splenocytes, a structure-activity relationship study. **Biochemical Pharmacology**, v. 76, n. 4, p. 495-506, 2008.

OLIVEIRA, D. M. C. de. **Screening five species of medicinal plants used in the Amazon by analysing histamine secretion**. 2013. 105 f. Thesis (Doctorate in Sciences) - Federal University of Pará, Belém, PA, 2013.

PARACAMPO, N. E. N. P. Connarus perrottetii var. angustifolius **Radlk.**

(Connaraceae): traditionally used as barbatimão in Pará, Belém, PA: Embrapa Amazônia Oriental, 2011.

PAREKH, J.; KARATHIA, N.; CHANDA, S. Evaluation of antibacterial activity and phytochemical analysis of Bauhinia variegate L. bark. **African Journal of Biomedical Research,** v. 9, p.53 -56, 2006.

PATEL, I. et al. Phytochemical studies on Mansoa alliacea (Lam.). **International Journal of Advances in Pharmaceutical Research**, v. 4, n. 6, p. 1823-1828, 2013.

PEREIRA, R. J.; CARDOSO, M. das G. Vegetable secondary metabolites and antioxidants benefits. **Journal of Biotechnology and Biodiversity**, v. 3, n. 4, p. 146152, 2012.

PETROVSKA, B. B. Historical review of medicinal plants' usage. **Review Pharmacognosy**, v. 6, n. 11, p. 1-5, 2012.

PIETTA, P. Flavonoids as antioxidants. **Journal of Natural Products**. v. 63, n. 7, p.1035-1042, 2000.

PORTO, da C.; PORRETTO, E.; DECORTI, D. Comparison of ultrasound-assisted extraction with conventional extraction methods of oil and polyphenols from grape (Vitis vinifera L.) seeds. **Ultrasonics Sonochemistry**, v. 20, n. 4, p. 1076-1080, 2013.

PORT'S, P.S.P. et al. The phenolic compounds and the antioxidant potential of infusion of herbs from the Brazilian Amazonian region. **Food Research International,** v. 53, n. 2, p. 875-881, 2013.

REZENDE, H. A.; COCCO, M. I. M. The use of phytotherapy in the daily life of a rural population. **Revista Escola de Enfermagem USP**, v. 36, n. 3, p. 282-288, 2002.

RIBEIRO, C. M. **Avaliação da atividade antimicrobiana de plantas utilizadas na medicina popular da Amazônia**. 2008. f. 66. Dissertation (Master's Degree in Pharmaceutical Sciences) - Universidade Federal do Pará, Belém, PA, 2008.

ROCHA, G. G. et al. Natural triterpenoids from Cecropia lyratiloba are cytotoxic to both sensitive and multidrug resistant leukaemia cell lines. **Bioinorganic & Medicinal Chemistry**, v. 15, n. 23, p. 7355-7360, 2007.

SANTOS, R. I. **Basic metabolism and origin of secondary metabolites**. In: SIMÕES, C. M. O. et al. Farmacognosia - da Planta ao Medicamento. 5. ed. Porto Alegre: Editora da UFRG/Editora da UFSC, 2004. ch. 16, p. 404, 411-412, 422.

SARJIT, A.; Yi, W.; GARY, A. D. Antimicrobial activity of gallic acid against thermophilic Campylobacter is strain specific and associated with a loss of calcium ions. **Food Microbiology**, v. 46, p. 227-233, 2015.

SCHMITZ, W.; SAITO, Y. A.; SARIDAKIS, H. O. D. Green tea and its actions as a chemoprotector. **Semina: Biological and Health Sciences,** v. 26, n. 2, p.119- 130, 2005.

SHAHIDI, F.; AMBIGAIPALAN, P. Phenolics and polyphenolics in foods, beverages and spices: Antioxidant activity and health effects: A review. **Journal of Functional Foods**, v. 18, p. 820-897, 2015.

SILVA, C. T.; JASIULIONIS, M. G. Relationship between oxidative stress, epigenetic alterations and cancer. **Ciência e Cultura**, v. 66, n. 1, p. 38-42, 2014.

SILVEIRA, G. D. et al. Determination of Phenolic Antioxidants in Amazonian Medicinal Plants by HPLC with Pulsed Amperometric Detection. **Journal of Liquid Chromatography & Related Technologies,** v. 38, p. 1259-1266, 2015.

STALIKAS, C. D. Extraction, separation, and detection methods for phenolic acids and flavonoids. **Journal Separation Science**, v.30, n. 18, p.3268-3295, 2007.

TIVERON, A. P. Antioxidant activity of Brazilian vegetables and its relation with phenolic composition. **International Journal of Molecular Sciences**, v. 13, n.7, p. 8943-8957, 2012.

VEIGA, V. F. J.; PINTO, A. C.; MACIEL, M. A. M. Plantas medicinais: cura segura. **Química Nova**, v. 28, n. 3, p. 519-528, 2005.

VILKHU, K. et al. Applications and opportunities for ultrasound assisted extraction in the food industry - A review. **Innovative Food Science and Emerging Technologies**, v. 9, n. 2, p. 161-169, 2008.

VINATORU, M. An overview of the ultrasonically assisted extraction of bioactive principles from herbs. **Ultrasonics Sonochemistry**, v. 8, p. 303-313, 2001.

WU, C. et al. A comparison of volatile fractions obtained from Lonicera macranthoides via different extraction processes: ultrasound, microwave, soxhlet extraction, hydrodistillation, and cold maceration. **Integrative Medicine Research**, v. 4, n. 3. p. 171-177, 2015.

YAO, L. H. et al. Flavonoids in food and their health benefits. **Plant Foods for Human Nutrition**. v. 59, p. 113-122, 2004.

ZOGHBI, M. G. B.; OLIVEIRA, J.; GUILHON, G.M. P. The genus Mansoa (Bignoniaceae): a source of organosulfur compounds. **Brazilian Journal of Pharmacognosy**, v. 19, n. 3, p. 795-804, 2009.

ZUANAZZI, J. A. S.; MONTANHA, J. A. **Flavonoids**. In: SIMÕES, C. M. O. et al. Pharmacognosy - from Plant to Drug. 5. ed. Porto Alegre: Editora da UFRG/Editora da UFSC, 2004. ch. 23, p. 577, 601-607.

CHAPTER 3

PHENOLIC CONSTITUENTS IN MEDICINAL SPECIES OF CONNARUS, CECROPIA AND MANSOA

Fernanda Brum Pires

Carolina Bolssoni Dolwitsch

Marcelo Barcellos da Rosa

The use of medicinal plants as a therapeutic resource for health prevention, promotion and recovery has been practised for thousands of years since the dawn of humanity (JUNIOR; PINTO; MACIEL, 2005; SCHMITZ et al., 2005; ALVIM et al., 2006). This activity arose from popular knowledge and for a long time was the only alternative treatment for health problems (ALVIM et al., 2006).

Throughout history, the therapeutic use of medicinal plants has undergone a change of behaviour, influenced by the prevailing culture of the time. However, in the 1980s and 1990s it was revived to act as a complement to modern, scientifically-based medicine (ALVIM et al., 2006).

In this scenario, the use of plants today continues to expand, given that, according to the World Health Organisation (WHO), around 80% of the world's population uses traditional medicine to meet their primary health care needs (BRASIL, 2006).

Brazil, as a country of noble plant biodiversity and holder of the largest block of green area on the planet, the Amazon Rainforest (PARACAMPO, 2011), has a significant use of plants for medicinal purposes. However, knowledge about the chemical composition of plants is very limited, as the vast majority are used with little or no scientific evidence (CAMPELO, 2006). In order to change this panorama, relevant studies involving the chemical constitution of plant species are necessary in order to substantiate their pharmacological properties.

It is known that the medicinal properties of plants result from the presence of secondary metabolites in their constitution (PAREKH; KARATHIA; CHANDA, 2006), including phenolic compounds, a group widely distributed in the plant kingdom that exert antioxidant, antibacterial, antiviral (SIMÕES et al., 2004), anti-inflammatory and antiallergic properties (PIETTA, 2000). They provide protection against ultraviolet radiation and attacks by pathogens such as insects, fungi, viruses and bacteria (SIMÕES et al., 2004; TIVERON et al., 2012).

In recent years, there has been a growing interest in both research and human consumption of plant polyphenols, due to their nutritional potential and therapeutic value (AJILA et al., 2011). However, numerous species of medicinal interest still need to be investigated in relation to this and other classes of compounds, including: Connarus perrottetti var. angustifolius, Cecropia palmata, Cecropia obtusa and Mansoa alliacea, native to the Amazon, extensively used by the local population and with little scientific information.

The species Connarus perrottetti var. angustifolius, known as barbatimão do Pará, belongs to the Connaraceae family and is widely used for its antidiarrhoeal, anti-haemorrhagic, anti-inflammatory, antibacterial, antifungal, antiviral and healing properties (PARACAMPO, 2011), as well as for treating genitourinary infections in women, uterine bleeding, vaginal discharge, headaches, gastric diseases and coughs (COELHO-FERREIRA, 2009).

The Cecropia palmata and Cecropia obtusa species belong to the Moraceae family, whose popular names are red and white Embaúba, respectively (HOMMA, 2003). The genus has several activities recognised by popular medicine, such as: antidiabetic (FREITAS; FERNANDES, 2006), expectorant, mucolytic, antiseptic, laxative, antimicrobial (LAMEIRA et al., 2006), diuretic, antitussive and anti-inflammatory (FREITAS; FERNANDES, 2006; LAMEIRA et al., 2006; COSTA et al., 2011).

In Brazil, especially in Pará, Mansoa alliacea is popularly known as "cipó-d'alho" due to the characteristic garlic smell of the leaves when macerated (ZOGHDI;

OLIVEIRA; GUILHON, 2009). Belonging to the Bignoniaceae family, it is used as an analgesic, antipyretic, antirheumatic, used to treat respiratory diseases (RIBEIRO, 2008), antimalarial (PERÉZ, 2002), antifungal and antiviral (ZOGHDI; OLIVEIRA; GUILHON, 2009).

Various factors can coordinate or alter the rate at which plants produce phenolic compounds and other secondary metabolites. In this way, the period in which a plant is collected is one of the most important factors that can interfere with the quantity of compounds, even though their nature may not remain constant throughout the year (GOBBO-NETO; LOPES, 2007).

Phenolic compounds can vary from simple structures to highly polymerised substances, which can also be complexed with various plant components, which is why different extraction methods and solvents with different polarities are required to obtain them (NACZK; SHAHIDI, 2004).

The aim of this study was to identify and quantify phenolic constituents such as gallic acid, catechin, caffeic acid, rutin, ferulic acid, quercitrin, myricetin, fisetin, resveratrol, quercetin, kaempferol, chrysin and flavone, present in Connarus perrottetti var. angustifolius, Cecropia palmata, Cecropia obtusa and Mansoa alliacea, collected at the EMBRAPA Amazônia Oriental Garden (Belém, PA), between 2012 and 2014, obtained by extraction with ethanol/water, n-butanol and ethyl acetate. To this end, high-performance liquid chromatography with diode array technology was used for separation, identification and quantification. In addition, we sought to establish a discussion involving the support and applicability of these species in medicine.

EXPERIMENTAL

Materials, reagents and solutions

The analytical standards gallic acid, (+)-catechin, caffeic acid, rutin, quercitrin, myricetin, fisetin, quercetin, kaempferol, chrysin and flavone were purchased from Sigma-Aldrich (St. Louis, MO, USA). Ferulic acid from Fluka (Buchs, Switzerland)

and Resveratrol (Tedia, Rio de Janeiro, RJ, Brazil). All the standards used are of analytical grade, with a minimum purity of 95%.

The 85% (v/v) PA phosphoric acid reagent (F. Maia, Brazil) used to prepare the mobile phase was diluted in ultrapure water, which was obtained from a Milli-Q system (Millipore Synergy UV, Bedford, MA, USA). After dilution in a volumetric flask to a concentration of 0.1% (m/m), the solution was filtered using a vacuum system (Prismatec, model 131, 2 V) with a 0.2 μm cellulose acetate membrane filter (Sartorius, Goettingen, Germany).

The solvent acetonitrile, the elution component, was obtained from Panreac ITW Companis (Germany) and the methanol used for the dilutions of standards and samples from Tedia (Rio de Janiero, RJ, Brazil), both HPLC grade.

The stock solutions of the standards were prepared in volumetric flasks in the unit (μmol/L) by diluting each phenolic compound in methanol (HPLC grade), all corresponding to the equivalent (1000 $mg.L^{-1}$) concentration. After preparation, they were stored in Falcon tubes and kept in a freezer at - 30°C for analysis. Solutions of intermediate concentrations of phenolic compounds were prepared by diluting the stock solutions in methanol.

The plant samples were weighed on a digital analytical balance (Shimadzu/auy 220) with ± 0.0001 g precision.

Plant material

The bark of *Connarus perrottetti var. angustifolius* Radlk (IAN registration no. 184393) and the leaves of *Cecropia palmata* Willd (IAN registration no. 185556), *Cecropia obtusa* Trécul (IAN registration no. 185555) and *Mansoa alliacea* (IAN registration no. 184394) were collected in April and May in 2012, 2013 and 2014 and supplied by the Brazilian Agricultural Research Corporation (EMBRAPA), Eastern Amazon, Belém/PA.

The species grown in the EMBRAPA garden were collected, identified and sent dried and ground. The geographical location (1° 27'21"S latitude and 48° 30'14"W longitude with an altitude of 10 m and an average annual temperature of 30 °C) corresponds to where the samples were collected.

According to Souza and Ambrizzi 2003, the climate of the Amazon region is characterised by a "dry period" (from July to October) and a "rainy period" (from December to May), with the months of June and November being the transition periods. It has a hot and humid climate with very small temperature gradients. Therefore, the sample collection periods in this study correspond to the rainy season in the Amazon.

Extraction procedure

The plant extracts were obtained using ultrasound-assisted extraction (Bandelin Sonorex Super RK 510 H). The procedure was carried out in glass tubes containing approximately 0.2 g of the samples, to which 10 ml of the solvents (70% hydroethanol, 100% butyl alcohol or butanol and 100% ethyl acetate) were added and left for 4 hours in an ultrasonic bath, without the use of temperature. At the end of this period, the supernatant was removed from the crude extract and filtered through a 0.22 μm membrane (Sorbline Tecnologie). For the butanolic and ethyl acetate extracts, the solvents from the supernatant portion were evaporated in a glass beaker in an oven at 40°C. The solvents were then filtered through a 0.22 μm membrane (Sorbline Tecnologie).

resuspended in methanol (HPLC grade), maintaining the initial concentration, and then filtered through a 0.22 μm membrane (Sorbline Tecnologie). The extracted samples were stored in a freezer at 30°C until the analysis procedure. For injections into the chromatograph, all samples were diluted to 1% in methanol (HPLC grade).

Preparing infusions

The infusions were prepared by weighing out 1.5 g of the dried plant and adding 50 ml of water at 90°C for 30 minutes. After the infusion period, they were first filtered through filter paper and then through a 0.22 μm membrane (Sorbline Tecnologie). They were stored in a freezer at - 40°C until the analysis procedure. Samples were injected at 1% and diluted in methanol (HPLC grade).

HPLC-DAD instrumentation

For the chromatographic analyses, a Knauer liquid chromatograph (Berlin, Germany) was used, consisting of a Smartline Pump 1000 module coupled to a Smartline Manager 5000, and a detector using ultraviolet (UV) spectra scanning photodiode array technology, model Smartline 2600, managed by ChromGate® software (Knauer), version 3.3.1. It has a manual injector with a volume of 20μL.

Photodiode array high-performance liquid chromatography conditions

The phenolic compounds were identified and quantified using the methodology developed by Lima (2013), whose analytical validation showed it to be a selective, linear, precise and accurate chromatographic method with a low detection and quantification limit.

A gradient separation system was used, with a C18 250 x 4.6 mm reverse phase column, 5μm particle size and a pre-column of the same nature coupled, both (Phenomenex). The mobile phase used in the separation was composed of orthophosphoric acid (0.1%, m/m) as solvent A, and acetonitrile as solvent B. The elution conditions were as follows: 90-80% A and 10-20% B (0-5min); 80-75% A and 20-25% B (5-35 min); 75-0% A and 25-100% B (35-55min). Mobile phase flow rate: 0.8 $ml.min^{-1}$ (0-35min); 0.8-1.0 $ml.min^{-1}$ (35-55 min). The chromatographic runs were carried out by monitoring the ambient temperature (21°C±2). The ultraviolet scanning wavelengths used for detection were: 220 nm, 254 nm, 320nm and 360nm.

The chromatographic analyses of the plant extracts were based on comparing the chromatographic signals of the extracts and the reference standards using retention times, UV-Vis spectra and the addition of three known concentrations of standard solutions to the samples.

The compounds of interest in this study were quantified using the standard addition method, which relates the concentrations of the standard added in (μmol/L) to the areas obtained in (mAU). According to Ribani et al., 2004, by extrapolating the line, the concentration in the analysed sample is determined by the point where it cuts the abscissa axis (y).

Table 1 shows the retention times of the antioxidants found in the plant species, the wavelength used for detection (nm), the limit of detection and the limit of quantification of the chromatographic separation using the HPLC-DAD method.

Table 1: Chromatographic separation data using the HPLC-DAD method

Phenolic compound	t_R (min)	λ (nm)	LOD (mgkg)[1]	LOQ (mgkg)[1]
1. gallic acid	4.40	220	0.7	1.0
2. (+)- Catechin	8.65	220	1.7	2.8
3. Caffeic acid	10.10	320	0.3	1.0
4. Rutin	12.45	254	0.6	2.4
5. Ferulic acid	14.75	320	0.5	1.0
6. Quercitrin	18.00	254	0.4	0.8
7. Resveratrol	28.45	320	0.4	0.8

t_R : retention time; λ: wavelength; LOD: Limit of Detection; LOQ: Limit of Quantification.

The limits of detection (LOD) and quantification (LQ) were determined using the signal-to-noise method, in accordance with INMETRO document DOQ-CGCRE-008 of 2010, where the area and standard deviation corresponding to the retention times of the compounds of interest were determined by injecting 7 replicates of the blank (matrix free of the compound of interest), in this case methanol (HPLC grade). Using the equations:

$LD = X + t_{(n-1)}.s$ e $LQ = X + 10.s$ where: : X = mean of the values of the whites; s = sample standard deviation of the whites; t = number of injections (7-1) = 6 degrees of freedom; For these degrees of freedom, the value of one-sided t, for 99% confidence is 3.143.

RESULTS AND DISCUSSION

The methodology used to determine phenolic constituents by HPLC-DAD allows the separation and identification of thirteen antioxidants, seven of which were detected in the plants under study, such as: gallic acid, catechin, caffeic acid, rutin, ferulic acid, quercitrin and resveratrol.

Phenolic components in plant extracts from different collection years determined by HPLC-DAD

Connarus perrottetti var. Angustifolius

Compound	**2012**	**2012**	**2012**	**2013**	**2013**	**2013**
	Hidroet.	But.	Acet.	Hidroet.	But.	Acet.
Gallic acid	-	65.3	79.3	439	170.7	142.5
Catechin	4558.5	1330.2	1752.8	3121.9	589.6	569.3
Caffeic Acid	-	-	-	-	13.7	24.8
Rutin	-	-	-	-	18	5.9
Ferulic acid	-	-	-	-	11.7	-
Quercitrin	-	-	-	5.3	3.9	8.3
Resveratrol	-	-	-	-	17.5	11.5

*Concentration (mg kg)$^{-1}$

**ethanol/water (Hidroet.); n-butanol (But); ethyl acetate (Acet.)

Compound	**2014**	**2014**	**2014**
	Hidroet.	But.	Acet.
Gallic acid	594.9	40.3	31.4
Catechin	1983	115.1	307.9
Caffeic Acid	-	11.5	-
Rutin	-	23	34.8
Ferulic acid	-	3.9	-
Quercitrin	-	11.5	-
Resveratrol	- 1 -1x	17.7	-

*Concentration (mg kg)$^{-1}$

Considering the results shown, it can be seen that among the species analysed, Connarus perrottetti var. angustifolius was the one with the greatest diversity of phenolic compounds, being the only one to contain all seven compounds determined in the plants in question.

Among the analytes investigated, catechin was the predominant antioxidant in this species, corresponding to approximately 80% when compared to the sum of all the other components detected. It can be seen that all the extracts referring to this compound showed a similar behaviour in terms of decreasing concentration between the 2012 and 2014 collections, but there was no correlation in terms of proportion.

Considering the results shown, it can be seen that among the species analysed, Connarus perrottetti var. angustifolius was the one with the greatest diversity of phenolic compounds, being the only one to contain all seven compounds determined in the plants in question.

The highest amounts of gallic acid and catechin were found in the hydroethanolic extract, which can be explained by the high polarity of these compounds and solvents, as the presence of 30 per cent water increases the polarity of ethanol, facilitating the extraction of the more polar constituents (STALIKAS, 2007).

The use of solvents with lower polarity, such as butanol and ethyl acetate, allowed the identification of other constituents with lower polarity than gallic acid and catechin. This observation can be substantiated by NACZK, 2006, who argues that the solubility of phenolic compounds is influenced by the chemical nature of the plant and the polarity of the extracting solvent, which is why various organic solvents and aqueous mixtures with alcohols are reported in the process of extracting these metabolites (AJILA et al., 2011; IGNAT; VOLF; POPA, 2013).

Using liquid chromatography with amperometric detection, Da Silveira and colleagues (2015) evaluated the 2012 collection of this species and also verified the presence of gallic acid and catechin in the same extraction solvents used here.

Caffeic acid and ferulic acid had higher concentrations in the sample collected in 2013, while rutin and quercitrin had higher concentrations in 2014.

Resveratrol did not vary significantly between the 2013 and 2014 collections in the butanolic extract, however, compared to ethyl acetate in the collection of
In 2013, there was a reduction of approximately 34 per cent, showing a lower affinity for this solvent. However, it wasn't even detected in the 2012 collection.

Among the compounds detected in the species, rutin and resveratrol stand out as therapeutic agents in inflammatory processes (BABU; LIU; GILBERT, 2013; ANTUS et al., 2015). According to Coutinho et al. 2009, rutin can exert activity through two mechanisms: modulation of the cyclooxygenase enzyme and inhibition of pro-inflammatory cytokines. The latter process is also attributed to resveratrol (LÓPEZ-POSADAS et al., 2008), which has also been reported to have an antibacterial action against strains of Helicobacter pylori (BROWN et al., 2009), findings which may explain the properties referred to the species.

Studies by Oliveira, 2013, demonstrated the antiallergic activity of extracts of *Connarus perrottetti var. angustifolius*, due to the antisecretory action of histamine, which was attributed to the presence of compounds such as catechin and flavonoids in the composition of the species, in line with the results of this study.

Cecropia obtusa

Compound	**2012**	**2012**	**2013**	**2013**	**2014**	**2014**
	Hidroet.	But.	Hidroet.	But.	Hidroet.	But.
Gallic acid	4.2	24.6	16.9	14.9	53.8	-
Catechin	-	-	-	161.7	-	-
Caffeic Acid	-	-	-	-	-	7.8

*Concentration $(mg\ kg)^{-1}$

In relation to gallic acid, the hydroethanolic extract of this species showed a significant increase of 92.2% in the samples collected between 2012 and 2014. These findings may be due to some stress suffered by the species, since various environmental factors reported in the literature, such as seasonality, temperature, water availability, ultraviolet radiation, nutrients, altitude, among others, can interfere with the biosynthesis of secondary metabolites (GOBBO-NETO; LOPES, 2007; MORAIS, 2009).

The compound with the highest concentration in the plant was catechin, found only in the butanolic extract collected in 2013. Caffeic acid was detected in the same solvent, but only in the 2014 sample.

It can be seen that the species, as well as having little variety of antioxidants in its constitution, showed very irregular behaviour throughout the collections analysed. This may be related to the plant's defence strategy against the influence of biotic and/or abiotic factors, as mentioned by MORAIS, 2009 and COUTINHO et al. 2009.

As for the gallic acid and catechin found in this species, there are various activities in the literature, including antimicrobial action against Escherichia coli (GOMES et al., 2014; SARJIT; WANG; DYKES, 2015) and some strains of Helicobacter pylori (GOMES et al., 2013), in line with the use described in folk medicine.

Cecropia palmata

Compound	2012	2012	2013	2013
	Hydroethanolic	Butanol	Butanol	Ethyl acetate
Gallic acid	76.2	13.4	13.5	-
Catechin	-	-	159.6	-
Caffeic Acid	-	-	-	14.9
Ferulic acid	-	-	1.5	-
Quercitrin	-	-	22.4	-
Resveratrol	-	-	6	-

*Concentration (mg kg $)^{-1}$

Compared to the species of the same genus analysed in this study, this plant showed a greater diversity of the compounds under study.

In the 2012 collection, the hydroethanolic extract extracted a higher concentration of gallic acid. The high polarity of this compound and the solvent may explain this result. However, no other analytes of interest were obtained under this condition in the collections analysed.

In the material from 2013, it was noted that the butanol solvent enabled the extraction of both gallic acid and other compounds with a lower polarity than this that had previously been identified in this species, including catechin, ferulic acid, quercitrin and resveratrol, as well as caffeic acid, which obtained a higher affinity in ethyl acetate.

Catechin was the majority compound for both species of Cecropia, although it was only detected in the butanol fraction in the 2013 samples. Comparing the amount of this antioxidant for the two species, there was little significant variation, with C. obtusa obtaining 1.3% more catechin than C. palmata.

Several studies involving species of the Cecropia genus can be found in the literature. Fanck et al. 1998, using thin layer chromatography, found phenolic acids and flavonoids in extracts of the leaves of C. hololeuca, C. pachystachaya and C. glaziovii. Also for C. glaziovii using liquid chromatography, Caicedo (2005) identified catechin and Arend et al. 2011, as well as Beringhs et al. 2015, detected caffeic acid; these findings may support the results presented here for the two Cecropia species investigated here.

Compared to Connarus perrottetti var. angustifolius, C. palmata only lacked rutin in its composition.

Among the compounds detected, studies have shown that quercitrin and ferulic acid have the ability to increase insulin secretion by the beta cells of the pancreas, due to their antioxidant power in reducing the oxidative stress caused by diabetes (SRINIVASAN; SUDHEER; MENON, 2007; BABUJANARTHANAM et al., 2011). This idea may support the popular use of C. palmata as an antidiabetic.

Its use in folk medicine as an antimicrobial may also be based on the presence of gallic and ferulic acid, as these are recognised for their antibacterial action against Escherichia coli, Pseudomonas aeruginosa, Staphylococcus aureus and Listeria monocytogenes (BORGES et al., 2013; SARJIT; WANG; DYKES, 2015). The use of

Connarus perrottetti var. angustifolius for this purpose can also be justified, as these compounds are common to both species.

Mansoa alliacea

Compound	2012	2012	2013	2014
	Hydroethanolic	Butanol	Butanol	Butanol
Gallic acid	13.4	-	-	-
Caffeic Acid	-	69.4	58.2	8.4

* Concentration (mg kg $)^{-1}$

This species had the lowest diversity of phenolic compounds. However, in relation to the species studied, it had the highest concentration of caffeic acid, detected in the butanolic extract. Between the 2012 and 2014 collection periods, there was an 87.9 per cent reduction in the amount of this compound.

Like the other vegetables evaluated, it also had gallic acid in its composition, but only in the samples collected in 2012.

Among the compounds of interest in this research, phytochemical studies involving the leaves of Mansoa alliacea have revealed the presence of polyphenols and flavonoids (RIBEIRO, 2008; PATEL et al., 2013) and the absence of catechins (RIBEIRO, 2008). The flowers contain flavonoids such as apigenin and luteolin (ZOGHBI et al., 2009), and the roots contain phenols (PATEL et al., 2013).

The analyses and results presented in this study add to our knowledge of the chemical composition of this species, as they confirm the absence of catechin already mentioned in the literature and also reveal phenolic acids such as caffeic and gallic acids in this plant.

According to Simões et al. 2004, caffeic acid plays an antioxidant role and its continued intake through the diet can delay diseases caused by oxidative stress.

According to the National Meteorological Institute (INMET), the average rainfall during the collection periods (2012, 2013 and 2014) was 416.7±80.4 nm,

with a coefficient of variation of around 20%, with minimum temperatures of 23.03±0.3 °C and maximum temperatures of 32.8±0.3 °C. According to Becho et al. (2009) the more polar compounds can be eliminated by plants through leaching, and considering that the samples involved in the study represented the rainy season in the Amazon, these findings may support the behaviour observed throughout the collection period.

Although the chemical constitution of plants is mainly determined by the genetic characteristics of the species, it is known that various environmental factors and different extraction methods can interfere with their composition (DINIZ; ASTARITA; SANTARÉM, 2006), so it is believed that each species can show a specific behaviour in the face of variations, which is why studies in this area are fundamental for plant quality control.

Analysis of infusions by HPLC-DAD

Among the analyses of the infusions of the plant species, it was possible to verify that only Connarus perrottetti var. angustifolius and Cecropia obtusa had the compounds researched. Relating the quantities detected to therapeutic use in the form of tea, Connarus perrottetti var. angustifolius obtained 0.81mg/100ml of gallic acid in the 2013 sample and 0.58mg/100ml in the 2014 sample. Catechin content was 7.9mg/100ml in the 2013 sample and 3.5mg/100ml in the 2014 sample.

The 2012 and 2013 samples of Cecropia obtusa showed no quantitative variation, with 0.14mg/100ml of gallic acid.

Considering its use for medicinal purposes, the consumption of tea obtained by infusing the bark of Connarus perrottetti var. angustifolius can provide reasonable amounts of active compounds for the body, such as gallic acid and catechin, which are associated with various pharmacological uses, mainly due to their recognised antioxidant activity.

CONCLUSION

The medicinal plants from the Amazon in this study had one or more phenolic compounds, including phenolic acids and flavonoids, in the chemical constitution of their extracts.

Considering the results obtained by diode array detection, phenolic, gallic and caffeic acids were the only antioxidants found in all the species studied.

In Connarus perrottetti var. angustifolius, catechin, rutin, ferulic acid, quercitrin and resveratrol were also found, of which only rutin was not found in Cecropia palmata. Catechin was only not detected in Mansoa alliacea.

However, in the species that were in the composition, it was the majority compound. The infusion of Connarus perrottetti var. angustifolius can provide significant amounts of gallic acid and catechin, supporting its use for medicinal purposes.

The phenolic compounds quantified in the medicinal species under study are known for their various pharmacological properties, such as antioxidant, antitumour, antimicrobial and antiallergic properties.

Although the phenolic constituents found in the species studied have significant support in the literature in terms of the pharmacological activities described by popular use, it should be noted that the results of this study show that the plants studied do not have the same constituents in different years of collection, nor even the same concentration.

This makes it clear that there is a need for detailed chemical analyses of plants intended for therapeutic use in order to guarantee the desired concentration of active ingredients.

It is also important to consider the importance of studies involving species belonging to the biodiversity of Brazilian flora, since they have a wide range of uses in folk medicine as well as nutritional appeal.

CHAPTER 4

FINAL CONSIDERATIONS

- The techniques used for detection and quantification, as well as the methodology for confirmatory analysis, made it possible to achieve the objectives set for the work.

- A chromatographic profile of the phenolic constituents of the species was obtained using diode array detection, which determined: gallic acid, catechin, caffeic acid, ferulic acid, rutin, quercitrin and resveratrol.

- Of the results obtained by HPLC-DAD, Connarus perrottetti var. angustifolius was the only species to have all the antioxidants described above; Cecropia palmata only lacked rutin; all the species had catechin as their main compound, with the exception of Mansoa alliacea, which was the only one not to have this constituent; all the plants had caffeic acid, which had a higher concentration in Mansoa alliacea, although it had a lower diversity of polyphenols in its constitution.

- Analyses of the infusions of Connarus perrottetti var. angustifolius provide significant amounts of gallic acid and catechin, supporting their use for therapeutic purposes in the form of tea, while Cecropia obtusa can also contribute gallic acid, although in much lower quantities than the former. It was not possible to determine the presence of polyphenols in the infusions of Cecropia palmata and Mansoa alliacea.

- The species in question showed differences in the composition and concentration of polyphenols over the period evaluated, however, various factors can influence the rate of production of secondary metabolites by plants, so future studies involving the monitoring of these parameters, such as rainfall, luminosity, seasonality, among others, are necessary in order to gain a greater understanding of behaviour.

- It was possible to relate the main activities mentioned by popular use to the presence of the phenolic compounds identified, thus concluding that the activities performed by the species may be the result of their presence in their constitution.

- The importance of studies involving the analysis of plant chemical composition is evident, in order to ensure that the species used in therapeutics have the ideal concentration of active ingredients to perform the required activities. It is also relevant because it covers plants from the Brazilian flora that are widely used by the population with little scientific proof.

REFERENCES

AJILA, C. M. et al. Extraction and analysis of polyphenols: recent trends. **Critical reviews in biotechnology**, v. 31, n. 3, p. 227-249, 2011.

ALVIM, N. A. T. et al. The use of medicinal plants as a therapeutic resource: from the influences of professional training to the ethical and legal implications of its applicability as an extension of the nurse's caring practice. **Revista Latino-americana de enfermagem**, v. 14, n. 3, p. 316-323, 2006.

ANTUS, C. et al. Anti-inflammatory effects of a triple-bond resveratrol analog: structure and function relationship. **European journal of pharmacology**, v. 748, p. 61-67, 2015.

AREND, D. P. **Development of a microstructured system containing a standardised extract of *Cecropia glaziovii* Sneth (Embaúba)**. 2010. 198 p. Dissertation (Master's in Pharmacy) - Federal University of Santa Catarina, Florianópolis, SC, 2010.

BABU, P. V. A.; LIU, D.; GILBERT, E. R. Recent advances in understanding the anti-diabetic actions of dietary flavonoids. **The Journal of nutritional biochemistry**, v. 24, n. 11, p. 1777-1789, 2013.

BABUJANARTHANAM, R. et al. Quercitrin a bioflavonoid improves the antioxidant status in streptozotocin: induced diabetic rat tissues. **Molecular and cellular biochemistry**, v. 358, n. 1-2, p. 121-129, 2011.

BECHO, J. R. M.; MACHADO, H.; GUERRA, M. O. Rutin-Structure, Metabolism And Pharmacological Potency. **Rev Int Est Exp**, v. 1, p. 21-25, 2009.

BERINGHS, A. O. et al. Response Surface Methodology IV-Optimal design applied to the performance improvement of an RP-HPLC-UV method for the quantification of phenolic acids in Cecropia glaziovii products. **Revista Brasileira de Farmacognosia**, v. 25, n. 5, p. 513-521, 2015.

BORGES, A. et al. Antibacterial activity and mode of action of ferulic and gallic acids against pathogenic bacteria. **Microbial Drug Resistance**, v. 19, n. 4, p. 256265, 2013.

BROWN, J. C. et al. Antibacterial effects of grape extracts on *Helicobacter pylori.* **Applied and environmental microbiology**, v. 75, n. 3, p. 848-852, 2009.

CAICEDO, P. E. L. **Contribution to the standardisation of Cecropia glaziovii Snethl leaf extracts: studies of seasonal and intraspecific variation in flavonoids and proanthocyanidins, extraction methodologies and vasorelaxant activity**. 2005.

CAMPELO, P. M. S. Medicinal plants and their extracts: the need for continued studies. **Estud. Biol**, v. 28, p. 62, 2006.

COELHO-FERREIRA, M. Medicinal knowledge and plant utilisation in an Amazonian coastal community of Marudá, Pará State (Brazil). **Journal of Ethnopharmacology**, v. 126, n. 1, p. 159-175, 2009.

COSTA, G. M. et al. An HPLC-DAD method to quantification of main phenolic compounds from leaves of Cecropia species. **Journal of the Brazilian Chemical Society**, v. 22, n. 6, p. 1096-1102, 2011.

COUTINHO, I. D. et al. Influence of seasonal variation on flavonoid content and antioxidant activity of *Campomanesia adamantium* (Cambess.) O. Berg, Myrtaceae leaves. **Revista Brasileira de Farmacognosia**, v. 20, n. 3, p. 322-327, 2010.

COUTINHO, M. A. S; MUZITANO, M. F.; COSTA, S. S. Flavonoids: Potential therapeutic agents for the inflammatory process. **Revista Virtual de Química**, v. 1, n. 3, p. 241-256, 2009.

DA SILVEIRA, G. D. da. et al. Determination of Phenolic Antioxidants in Amazonian Medicinal Plants by HPLC with Pulsed Amperometric Detection. **Journal of Liquid Chromatography & Related Technologies**, v. 38, n. 13, p. 1259-1266, 2015.

DE MORAIS, L. A. S. Influence of abiotic factors on the chemical composition of essential oils. **Hortic. bras**, v. 27, n. 2, 2009.

DÍAZ-GÓMEZ, R. et al. Comparative antibacterial effect of gallic acid and catechin against *Helicobacter pylori.* **LWT-Food Science and Technology**, v. 54, n. 2, p. 331-335, 2013.

DINIZ, A. C. B.; ASTARITA, L. V.; SANTARÉM, E. R. Alteration of secondary metabolites in *Hypericum perforatum* L. (Hypericaceae) plants submitted to drying and freezing. **Acta Bot Bras**, v. 21, p. 442-450, 2007.

FACCIN, H. et al. Study of ion suppression for phenolic compounds in medicinal plant extracts using liquid chromatography-electrospray tandem mass spectrometry. **Journal of Chromatography A**, v. 1427, p. 111-124, 2016.

FRANCK, U. **Phytochemische und pharmakologische Untersuchungen der kardiovaskulären Wirkprinzipien von Cecropia hololeuca Miq., Cercropia pachystachya Tréc., Cecropia glaziovii Sneth., Musanga cecropioides R. Brown und Crataegus L. monogyna, oxyacantha**. GCA-Verlag, 1998.

FREITAS, J. C. de; FERNANDES, M. E. B. Use of medicinal plants by the community of Enfarrusca, Bragança, Pará. **Boletim do Museu Paraense Emílio Goeldi Ciências Naturais**, v. 1, n. 3, p. 11-26, 2006.

GOBBO-NETO, L.; LOPES, N. P. Medicinal plants: factors influencing the content of secondary metabolites. **Química nova**, v. 30, n. 2, p. 374, 2007.

HOMMA OYAMA, A. K. **History of agriculture in Amazonia. From the pre-Columbian era to the third millennium**. EMBRAPA Amazonia Oriental, Bélem, PA (Brazil), 2003.

NATIONAL METEOROLOGICAL INSTITUTE, 2015. **http://www.inmet.gov.br/sim/gera graficos.php,** accessed in October 2015.

INSTITUTO NACIONAL DE METROLOGIA, **Normalização e Qualidade Industrial (INMETRO)**; Orientações sobre Validação de Métodos Analíticos, DOQ-CGCRE-008, 2010.

JUNIOR, V. F.; PINTO, A. C.; MACIEL, M. A. M. Plantas medicinais: cura segura. **Química nova**, v. 28, n. 3, p. 519-528, 2005.

LAMEIRA, O. A. et al. Phenology and Phytochemical Analysis of Medicinal Plants Occurring in Amazonia. **Embrapa Amazônia Oriental**, 2006.

LIMA, F. O. Comparative study of the bioactivity of phenolic compounds in medicinal plants. 2013. Thesis (Doctorate in Chemistry) - Federal University of Santa Maria, Santa Maria.

LÓPEZ-POSADAS, R. et al. Effect of flavonoids on rat splenocytes, a structure-activity relationship study. **Biochemical pharmacology**, v. 76, n. 4, p. 495-506, 2008.

MATSUMURA, E. et al. **Natural Products: Phytochemistry, Botany and Metabolism of Alkaloids, Phenolics and Terpenes**. 2013.

MS - Ministério da Saúde: **Política Nacional de Plantas Medicinais e Medicamento Fitoterápicos,** 1th ed.; Brasília: Brasil, 2006.

NACZK, M.; SHAHIDI, F. Extraction and analysis of phenolics in food. **Journal of Chromatography A**, v. 1054, n. 1, p. 95-111, 2004.

NACZK, M.; SHAHIDI, F. Phenolics in cereals, fruits and vegetables: Occurrence, extraction and analysis. **Journal of pharmaceutical and biomedical analysis**, v. 41, n. 5, p. 1523-1542, 2006.

OLIVEIRA, D. M. C. et al. **Screening of five medicinal plant species used in the Amazon by analysing histamine secretion**. 2013.

PARACAMPO, N. E. N. P. Connarus perrottetii var. angustifolius Radlk.(Connaraceae): traditionally used as barbatimão in Pará. **Infoteca Embrapa**, 2011.

PAREKH, J.; KARATHIA, N.; CHANDA, S. Evaluation of antibacterial activity and phytochemical analysis of *Bauhinia variegata* L. bark. 2006.

PATEL, I. Phytochemical studies on *Mansoa alliacea* (Lam.). **International Journal of Advances in Pharmaceutical Research**, v. 4, n. 6, p. 1823-1828, 2013.

PEREIRA, C. G.; MEIRELES, M. A. A. Supercritical fluid extraction of bioactive compounds: fundamentals, applications and economic perspectives. **Food and Bioprocess Technology**, v. 3, n. 3, p. 340-372, 2010.

PÉREZ, D. Etnobotánica medicinal y biocidas para malaria en la región Ucayali. **Folia amazónica**, v. 13, n. 1-2, p. 87-108, 2006.

PIETTA, P. Flavonoids as antioxidants. **Journal of natural products**, v. 63, n. 7, p. 1035-1042, 2000.

RIBANI, M. et al. Validation of chromatographic and electrophoretic methods. **Química nova**, v. 27, p. 771-780, 2004.

RIBEIRO, C. M. et al. **Evaluation of the antimicrobial activity of plants used in Amazonian folk medicine**. 2008. Master's dissertation. Federal University of Pará.

SCHMITZ, W. et al. Green tea and its actions as a chemoprotector. **Semina: Biological and Health Sciences**, v. 26, n. 2, p. 119-130, 2005.

SIMÕES, C. M. O. et al. **Pharmacognosy: From Natural Product to Medicine**. Artmed Editora, 2016.

SOUZA, E.; AMBRIZZI, T. Pentad precipitation climatology over Brazil and the associated atmospheric mechanisms. **Climanálise, January,** 2003.

SRINIVASAN, M.; SUDHEER, A. R.; MENON, V. P. Ferulic acid: therapeutic potential through its antioxidant property. **Journal of Clinical Biochemistry and Nutrition**, v. 40, n. 2, p. 92-100, 2007.

STALIKAS, Constantine D. Extraction, separation, and detection methods for phenolic acids and flavonoids. **Journal of separation science**, v. 30, n. 18, p. 32683295, 2007.

TIVERON, A. P. et al. Antioxidant activity of Brazilian vegetables and its relation with phenolic composition. **International journal of molecular Sciences**, v. 13, n. 7, p. 8943-8957, 2012.

ZOGHBI, M. G. B.; OLIVEIRA, J.; GUILHON, G. M. S. P. The genus Mansoa (Bignoniaceae): a source of organosulfur compounds. **Revista Brasileira de Farmacognosia**, v. 19, n. 3, p. 795-804, 2009.

Printed by Books on Demand GmbH, Norderstedt / Germany